"Man becomes aware of his own personality when he
seeks to come as close as possible to nature."

TEILHARD DE CHARDIN

Native Trees of Brazil

CECILIA BEATRIZ DA VEIGA SOARES

Native Trees of Brazil

Introduction
LUIZ EMYGDIO DE MELLO FILHO

SALAMANDRA

Editorial Project
MARCOS DA VEIGA PEREIRA

Design
LUIZ SERGIO BITTENCOURT
NARA ABUD TAUILE

Illustrations
DULCE TEREZINHA DO NASCIMENTO

Desktop Publish
HENRIQUE GARCIA DE LIMA

Version
ANNETTE BAUGHAN

Copy-desk
LIANA FORTES

Photos
Carlos Terrana/Kino Fotoarquivo – pages 81 and 91
Explorer/Sigla Fotoarquivo – page 67
Fábio C. Medeiros/Kino Fotoarquivo – page 93
Fotoarquivo Jorge's Estúdio – cover, pages 3 and 100
Haroldo Palo Jr./Kino Fotoarquivo – pages 37 and 42
José Carlos da Veiga Soares – pages 52, 53 and 57
Marco A. Fontes de Sá/Kino Fotoarquivo – page 64
Paula Laclette – page 84
Regina Lustosa – pages 38, 40, 41, 44, 45, 48, 58, 59, 60, 62, 63, 66, 68, 69, 70, 72, 73, 76, 77, 78, 83, 85, 86, 87, 92, 96, 99 and 109
Ronaldo Kotscho/Kino Fotoarquivo – page 80
Silvestre P. Silva/Angular Fotoarquivo – pages 43, 46, 47, 48, 49, 51, 54, 55, 56, 60, 61, 63, 65, 71, 74, 75, 79, 82, 88, 89, 90, 94, 98, 99, 101, 102, 103, 104, 105 and 106/107
Syd Lucas/Casa da Foto – pages 41, 50 and 95
Zig Koch/Kino Fotoarquivo – page 36
Zig Kock – pages 12/13, 39, 42, 97 and 108

Sindicato Nacional dos Editores de Livros, RJ. CIP-Brasil Catalogação-na-fonte

S653a Soares, Cecilia Beatriz Levy da Veiga
Árvores Nativas do Brasil / Cecilia Beatriz da Veiga Soares.— Rio de Janeiro: Salamandra, 1990.

ISBN 85-281-0007-3

1. Árvores—Brasil. I. Título.

90-0702

CDD - 634.970981
CDU - 634.0(81)

Salamandra Consultoria Editorial S.A.
Av. Nilo Peçanha, 155/301 — Phone: (021) 240-6306
20.020 – Rio de Janeiro – RJ
Brasil

Contents

Foreword

A new way, so to speak, of looking at all the events taking place on Earth. A new collective focus of human awarenesses may perhaps be the most prominent feature of the coming transition between the 20th and 21st centuries.

And this shift of coordinates of thinking is inclining toward a reappraisal of relations between man and the environment, as he recognizes his true role, no longer as the lord and master of a Nature at the service of his interests and ambitions, but, at most, as a capable, informed manager, caring for the "ad perpetuum" preservation of natural resources.

We are seeing the enlistment of increasingly numerous social groups, in the search for a deeper and more objective knowledge of the intricate mechanisms of the way Nature works and of the complex balances between her varying ecosystems. Everywhere and in all human communities, we are witnessing an outburst of voices of people anxious to halt the process of disorderly use of natural assets, or rather, the plundering of Nature.

In Brazil's case, with its exuberant Biota, whose wanton destruction has been going on for almost 500 years, since its discovery until now, almost half its natural forests have already been destroyed.

In the Atlantic Forest System, for example, which ranged from underwater vegetation to the misty woodlands along the crests of its forested mountains, the destruction has been almost total. What remains, 5% or less, consists of intermittent patches, which are now remnants to be protected at all costs. In this protection, the forces of the Government which defends, educates and punishes, and the communities which oversee, should be unified to intervene according to their possibilities, denouncing aggression and abuse, and urging the proper authorities to take action, even giving their support when official action is questioned or results in controversies.

In a country like ours, the home of some of the most luxuriant flora of any one nation, and as such the custodian of a true treasure of remarkable trees and bushes, it is amazing to find how few of them are made use of today.

It appears that what perhaps is lacking, is our authorities' enlightenment, expressed through encouragement given to those who are capable of doing the scholarly work of recognizing plants to be chosen in natural associations, of judiciously noting their characteristics, of collecting their seeds at the right times and in the right places, of establishing methods and techniques for their propagation, of domesticating them, and lastly, ofindicating the uses for which they are suitable.

Urban and regional landscaping, in the future, will be like a tool for ensuring the climatic comfort and aesthetical motivation of the generations that will replace us and guide the fate of the Earth in the next century.

This is the context of this book by Cecília Beatriz da Veiga Soares, whose enhanced sensitivity led her a long time ago to seek a closer relationship with the ornamental aspects of our flora. She has been an extraordinary cultivator of trees notable for their blossom, size, external appearance, coloring or any other attribute which valorizes them for specific uses in the area of landscaping projects, among them urban tree planting and ecological reforestation.

The methodology used, merging image and explanatory text, is enough to answer the questions of non-professionals and beginners alike and provides a wealth of useful information for the scholar and suggestions for the professional.

It is to be hoped that future contributions, along the same

lines, may offer information on other species as remarkable as those described here, and may arouse in other enthusiasts the desire to also write books that valorize and offer readers a broader range of knowledge about our botanical wealth.

However, returning to the subject of coming generations, I feel it is time to reflect a little and to make a real contribution toward improving the conditions prevailing during their time spent on this blue planet.

Cecília Beatriz's book gives the directions.

Rio, October 27, 1990

Luiz Emygdio de Mello Filho
Botanist

Introduction

I feel really privileged, because, since I was born, I have been accustomed to love, appreciate and respect nature. As I gradually became familiar with the plant world, I discovered a fascinating magical universe. It is a learning process that never ends. It seems that the more I try to know it, the more I realize how little I know, and the more I want to discover it.

At each moment I am surprised by incredible works of art in the delicacy of flowers, shapes, textures, in the most varied hues and very often in the different shades of the same leaf, like true artist's brush strokes.

It is rewarding when we learn to perceive all the beauty of a tree silhouetted against the twilight, observing each of its branches and their shapes, and feeling the intensity of the light illuminating its leaves. Another unique spectacle is the first blossoming of an Ipê-branco (Tecoma) early in the morning: tenuous sunbeams light up flowers still wet from the night's dew, making them shine.

This sort of reward is unique when one loves nature, and respects and works with her. At my country place, I have increasingly more proof of this: the more I plant, the more birds appear. A rare bird, entirely red, known as the Tié-sangue (Saddle Tanager), came the other day to greet me over breakfast.

Whenever there are fruit, a cobalt blue Tanager suddenly appears in a small bush near the window. Sometimes it perches on the orange-yellow flower of a creeper. A diminutive house wren once nested in the bedroom light fitting, while a hummingbird built its house on the electric cord of the table lamp.

A couple of mini-hummingbirds made a nest on the veranda, in a maidenhair fern pot, two meters from my eyes. They collected bits of fiber from the trunk of the "samambaiaçu" (a Dicksonia fern) from one end of the garden. At the other end, they pecked flakes of floss from high up the silk-cotton tree, constantly coming and going. I even saw them arranging the nest to make it comfortable.

One of the most remarkable happenings was a "bird convention", a very large number of birds, of an immense variety, spread out through various trees and bushes of the tree nursery. I was thus able to listen to a true concert, with a remarkable chorus.

Each of these moments is an unforgettable present. One cannot buy them and also there is no risk of losing them. They are with us forever.

We live in a country extremely well endowed by nature — we have one of the world's most varied collections of plant life, with very diversified vegetation. Most Brazilians, however, regardless of their standard of education or social level, are unaware of this wealth.

Urged by the almost total ignorance about our trees, I decided to write a book without any great pretensions, but which might contribute toward making people appreciate our flora.

Almost everyone, with very rare exceptions, calls any tree with yellow flowers an Ipê, although we have countless cassias and other species with yellow blossom at the most varied times of the year.

Few also know how to appreciate the trees with "fallen" leaves, that is, trees that lose their leaves, because they say that they dirty the ground. Before losing them, however, they give us a show of uncommon beauty, with their orange, red and copper coloring. In cold countries, the fall is considered one of the most beautiful seasons, precisely because of the trees with "fallen" leaves.

It is said that we have approximately 50 thousand classified plant species and there are still so many to be discovered and much food for study. Only 70 native species were chosen for this book.

Some were chosen because of their formation and shape and others because they blossom at different times of the year.

The possibility of having flowers the whole year round — the flowering varies according to the climatic changes and the location of the plants — is yet another gift of our land. In the

summer months we have the purple and pink hues of the *Tibouchina* (Quaresmeira, or Glorybushes), interspersed with the stronger yellow of the *Cassia occidentalis* (Fedegoso, or Coffee Senna) and the paler and softer *Cassia multijuga* (Aleluia, another Senna). Then comes the exquisite pink blossom of the *Chorisia* (Paineira or floss-silk tree). A little later on, the *Cordia superba* (grão-de-galo) which flowers profusely part of the year, covering itself with delicate white flowers, forming a carpet around itself. Soon after is the time of the *Triplaris* (formigueira or tachi), tall, slender ant trees, looking like flowering torches with their pink and red clusters.

The cold wind in winter takes it upon itself to strip the leaves of the *Erythrina* (mulungu, suinã, a coralbean) and the *Tabebuia* (Ipê) and, almost suddenly, the clusters of coral-red flowers of the *Erythrina falcata* begin to open up, attracting Maracanãs (birds very similar and a little smaller than macaws), always in groups of ten or twelve. The next to bloom is *Erythrina verna*, which covers itself with spectacular blood-red blossom, providing a lovely and exotic contrast to the vivid, shining green of noisy Maritacas (Pionus parrots) who arrive in flocks and perch on its branches, enjoying its flowers.

Nature continues to reward us, covering the *Tabebuia* (Ipê) with golden-yellow, white and purple mantles. From September onward, we have the violet-blue blossom of the *Jacaranda mimosifolia* (Jacarandá-mimoso, or sharpleaf jacaranda), the yellow of the *Schizolobium parahyba* (guapuruvu, of the Senna family) and of the *Caesalpinia* (Sibipiruna) and the delicate pink flowers of the *Cassia grandis*, resembling more an immense peach tree.

The survival of the human race is closely interrelated to the balance of this plant world. The fate of trees depends on man's will and, without any doubt at all, man's fate depends on the future of trees. The planet Earth is beginning to warn us that we must stop destroying. It is time to plant.

The story of the Curupira, a native myth which, according to legend, threw hunters off the scent to defend the forest, is an example to be followed. We urgently need several Curupiras scattered all over our woodlands. I would like many of us, very many, and increasingly more of us, to unite together, transformed into Curupiras, fighting for the preservation of what remains of our forests.

The Legend of the Curupira

Curupira is a Tupi myth which is part of our folklore. The legend has it that he is a smart boy, a mysterious genius, with many powers, protector of plants and animals. Dwelling in the heart of forests, he makes his home in the hollow trunks of big trees and appears always mounted on peccaries, with his ax made of turtle-shell.

His feet are turned backwards, with their heels facing forward, allowing him to trick and divert robbers of bird's nests, those who kill the young of any animal and who mutilate plants and creatures just for the pleasure of destroying.

Irritated by this destruction, the Curupira takes vengeance in his own way. He imitates all the four-legged animals and birds to lure the hunter, who follows him, imagining he is pursuing prey.

Sometimes he enchants men with the following song:

Cha uatá, uatá
Ce Rapé Rupi
Cha Uatá, Uatá
Ce Racacuera
Yure Uatá, Uatá, which means:

"I'm walking along my path, come along after me walking, walking".

And the hunter, led by the song, disappears into the woods until the Curupira stops singing and he realizes he is lost.

The Curupira is the symbol of the Brazilian Foundation for the Conservation of Nature (FBCN), a member of the International Union for the Conservation of Nature.

The Importance of Trees in Soil Conservation

Trees are essential for protecting the environment: air, soil, fauna, flora, water and climate. Where the land slopes above certain angles, the presence of trees prevents soil erosion, maintains and trains the course of streams, protecting the spring and the bed of rivers, shores of lakes and weirs. It also makes the climate milder, maintaining the relative humidity of the atmosphere.

As they grow, they enrich the soil, increasing its ability to absorb and store water. The leaves, dry branches, old trunks and other plant debris falling on the ground, decompose gradually and, helped by various agents, such as water and microorganisms, are transformed into humus, making this upper layer of soil extremely fertile.

The dense crowns of the trees attenuate the force of the rain, which then penetrates gradually into the soft, absorbent soil of the forest. When the rain falls in open areas, where the water sources are unprotected, the water flows down the slopes, causing sudden overflowing of streams, carrying an enormous amount of rich soil with it. In its path it leaves behind bare, unproductive earth, which in time becomes desert.

The Sahara and Brazil's Northeast are good examples: several studies have shown that they were once covered by thick vegetation. At the moment, we are seeing the fast desertification of the southern states of Brazil, affected by heavy floods and countless droughts.

The Tijuca Forest

Just to have an idea of the enormous importance of a forest, one should know something about the history of the Tijuca Forest. A large part of the population certainly thinks it has always been there. But this is not so.

As far back as the beginning of the last century, due to the rising population, Rio de Janeiro began to have serious water supply problems. The decreasing volume of the streams was put down to the destruction of the forests. Really, there had been total, devastating and irrational land clearing — everything had been turned into firewood and charcoal.

A report of that time describes the felling and clearing of the woodlands:

"The most valuable hardwoods were burned in formidable cutting and burning which consumed everything. Never had the world seen such a complete waste of such valuable flora, devoured in maelstroms of smoke and flames. Man laid waste the land in his feverish desire to expand his coffee plantations. The rows of Negroes, wielding axes, pushed the forest back toward inaccessible summits, to the stony and unusable crests."

The situation deteriorated with the arrival of Dom João VI and his immense entourage of more than 20 thousand noblemen and escorts. The neighboring woodlands were entirely devastated to meet the increased firewood consumption.

Between 1817 and 1818 the Government decided to take stringent action to prevent land clearing, ordering new preventive measures and punishments. However, as has happened in our country until today, the orders were never carried out. The city became more and more ravaged by drought. Attempts to protect the headwaters of rivers by replanting around them, did not produce results.

Only on December 11, 1861, did Dom Pedro II issue "provisional instructions" on the planting and conservation of the Tijuca Forest. The appointment of Major Manuel Gomes Archer as its administrator, marks the birth of the Tijuca Forest as we know it today. Reforestation began less than a month later. A herculean task requiring long treks on muleback, without qualified staff or technical resources. In spite of all these drawbacks the efforts were successful.

In the first year alone, 13,613 species were planted, while 10,215 were saved, including palms, jacarandas, mahogany, sapucaias, arcos de pipa, laurels, copaiba copals, oitis, myrtles, maçarandubas, spurges, jackfruit trees, muricis, pricklyashes, bamboos and many others. In a little over ten years, approximately one hundred thousand trees had been planted.

Major Archer was the precursor of forestry in Brazil, besides having been a pioneer in introducing it with mixed species. The planting of new seedlings, as well as the conservation and protection of the Forest, brought about a natural regeneration of species which had managed to survive the plundering. These factors as a whole contributed toward our being able today to enjoy the exuberant and lovely Tijuca Forest.

Information taken from the book *Floresta da Tijuca*, published by the Centro de Conservação da Natureza, Rio de Janeiro, 1966.

The Growth of Trees

Our vegetation is permanent and very varied — we have cold regions in the south, extremely hot, damp regions in the north, arid regions in the northeast and even other temperate regions.

Due to this privileged climate, trees grow much faster than in Northern Hemisphere countries. At 15 years of age, our Paraná Pine forests are already providing pulp for paper making. In Canada a pine takes a century to grow, in Finland 120 years and in certain regions of Russia, up to 170 years.

The *Ceiba pentandra* (sumaúma or silk-cotton tree), the largest tree in Amazônia, which the Indians call the "mother of trees", at 13 years old can be 22 meters high. Its trunk may reach one meter in diameter.

Very many species, such as the *Erythrina* (mulungo, or coralbean) can be planted with pieces of trunks or branches up to two meters high, without affecting their growth

Some months after sprouting, the seedling is already an adult tree.

Trees also reproduce through seeds which are carried by the wind or by birds, sometimes for very long distances. This is the case of *Schyzolobium parahyba* (bacarobu or guapuruvu), which reaches a height of two or three meters in less than a year. At eight years old, it may already be 12.5 m high and have a diameter of 21 cm. At 20, it is an enormous tree, 50 meters high with a majestic trunk up to 80 cm wide. Long before this, at five or six years old, it is ready for felling and for preparing lumber.

Very fast growing species like these, make it possible to plant woodlands or a forest in just a few years, provided man does not interfere with an ax, chain saw or cutting and burning. Nature's prompt reply is another good reason for planting a lot and increasingly more. The examples mentioned here are just some of those included in our flora, which is so little known and studied.

The Tree and Its Use in the Most Varied Sectors

Apart from its great beauty, the forest has a vital role to play in the development of the human race. Its best known product is lumber: lumber is used for building our homes and also for making furniture, instruments and countless objects. The forest also plays a fundamental part in the survival of birds and animals, guaranteeing their food and shelter.

Every part of a tree can be used: roots, trunks, leaves, flowers and fruit can be utilized fully for making oils, syrups, rubber, paper, charcoal, dyes, resins and various other products, in medicine, trade, chemicals and industry.

The possibilities of making use of the forest do not end there. The forest still has many resources which have been untouched until now. The food values of its species are practically unknown. The *Pereskia grandifolia* Haw (ora-pro-nóbis), a native of our country, is an example: its leaves are edible and can be prepared in the same way as spinach, collards or any other greens.

Because of lack of incentives for serious and continuous research, the potential of our plants as suppliers of raw materials for medicines has been almost entirely overlooked. The result is that we end by importing from multinationals medicines made from species of our own flora.

The preservation of forests, and above all the Amazon forest, is essential not only for the equilibrium of the entire ecosystem, but also to avoid the destruction of very useful plants that are still unknown. Unfortunately, man has been destroying, in an inconsistent and unproductive way, dozens,hundreds and even thousands of trees and does not realize that by destroying them, he is destroying himself as well.

Two simple measures would help maintain and replace the green that is essential for our life: the replacement of each felled tree with another and planting a seedling for each child born.

Then, probably, we would be able to attain the objective of one 40 hectare park for every 50 thousand inhabitants, which is considered ideal by specialists. The World Health Organization establishes a minimum of 12 m^2 of free area per inhabitant. In most cities these rates are not respected.

Modern urban centers represent for man not only the highest conquest in engineering and architectural methods, but also a serious problem. This is due partly to lack of suitable planning, so that the growth of such cities would be compatible with preservation of the environment.

Choosing a Suitable Tree

Choosing a tree can be entertaining and even agreeable. However, if good results are to be achieved, one must not only consider its beautiful appearance — foliage, shapes, flowers and their fragrance, textures and colors. The choice of a single plant as much as large landscaping projects — for city squares, parks, highways, factories and rural zones — require one to know a little about the origin and habits of each plant.

Two factors are extremely important: where it will be planted and the characteristics of the species. Without sufficient knowledge, cold climate trees may be transplanted to near the sea, exposed to the salty air and strong southwesterly winds, while others, natives of hot, humid areas, may end up in cold regions where they are exposed to frost. It should not be forgotten that plants are living beings which suffer when changes occur in their environment.

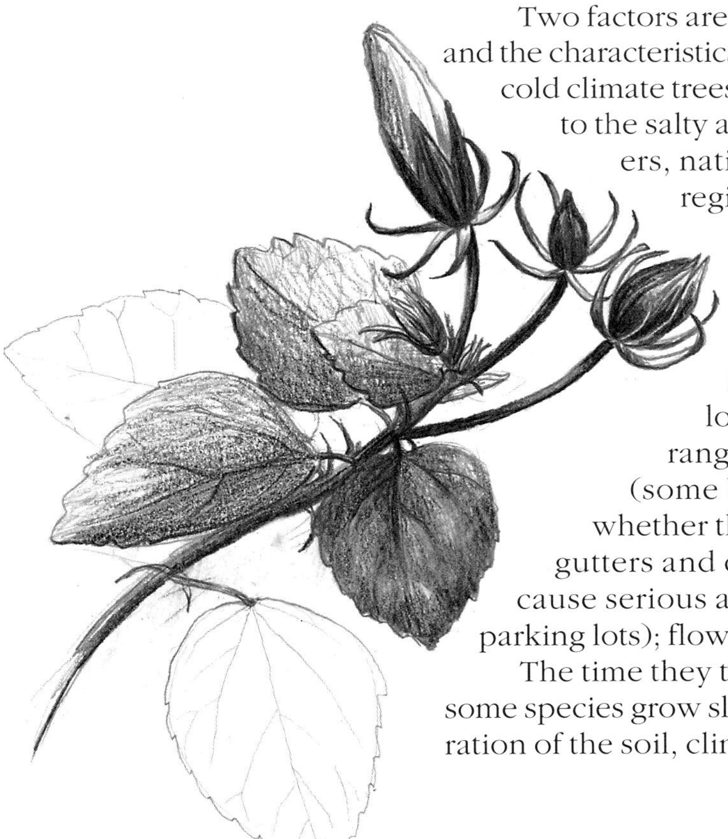

Certain basic notions are mandatory: climate, type of soil, root system (deep or shallow roots); trunk, height, shape, crown, the arrangement of branches, the branches themselves (some break easily); evergreen or "falling" leaves, whether they are large or small (small leaves can block gutters and drains); fruit (heavy fruit which fall and can cause serious accidents, or which stain, are not suitable for parking lots); flowering (coloring, blossom time and fragrance). The time they take to grow must also be taken into account: some species grow slowly, others faster. In certain cases, the preparation of the soil, climate and treatment they receive can speed up this process. Resistance to pests and blights must also be considered.

In hot, very sunny regions, leafy trees giving good shade are the best. In damp, cold places, the ideal is to have species which shed their leaves during the winter. As a shelter against strong winds, the trees should be planted perpendicularly toward the prevailing winds. In dry regions, drought-resistant species should be chosen.

Wide expanses of land call for large trees. Similarly, a very large tree should not be planted in a small space. On so many occasions, and this happens very often, people do not have the patience to wait for a plant to grow. In their haste, they do not foresee its future growth, nor do they allow it enough space, planting four to five seedlings in a place only big enough for one. The result is simply disastrous, because none of them may grow satisfactorily.

Whatever type of scheme is used, one should never disregard the species that most attract birds, providing them with food and shelter. It is recommendable also to use the largest possible number of native trees, already adapted to the environment.

In rural zones, trees are extremely useful for providing shade in pastures, coffee and cocoa plantations. Planted along fences and borders they are an economical and permanent protection, requiring neither maintenance nor periodical repairs. The wood can be used for countless purposes.

Urban tree planting contributes decisively to ecological balance, directly influencing the well-being of the city's inhabitants: it improves the quality of the air, narrows the temperature range, muffles noises, helps remove dust from streets and even provides new areas for leisure in woodlands, squares and areas for relaxation and recreation.

Various factors such as pollution tolerance, topography, the width of roads and avenues, open spaces, existing constructions, squares and parks must be considered in city planning, so that the vegetation may always be in harmony with the city.

Killer Trees

In the Forest, Trees Fight for Survival

For most people, the forest means just a group of varying trees. It is far more than this: it is an extremely well organized community of animals and plants, living in close association and varying degrees of interdependence. The struggle for satisfying basic needs is vigorous and only the fittest survive.

In this dispute, large trees avail themselves of all the means they have at their disposal. Some put out roots below the ground and gradually wrap themselves around the roots of a neighboring tree until they suffocate it. After it dies, the dominated tree falls over onto the other side, making room for the stronger tree to grow to its full splendor.

There are also species known as "mata-pau" (strangler trees). The most common examples are the "figueira" and the "clúsia" (*Clusia*). Starting with a seed carried by the wind or by birds, they fasten themselves to the branches of a large tree or to a palm trunk.

At first they grow very slowly. As soon as their roots reach the ground they gain incredible strength, and become wholly independent. Then begins the process of strangling the tree which sheltered them, enfolding it entirely.

For some time it is possible to single out the two trees because of their different foliage. In the end, this is practically impossible, unless one cuts the trunk. This is what we call a "perfect crime". There are also murderers in the plant world.

Guidance for Planting

When choosing trees for planting, seedlings should be selected that are free from pests or blights, are in good shape and preferably recently grown, because older specimens take a long time to recover from transplanting. The ideal height is between 1.80 m and 2.2 m, chiefly when public areas are to be planted.

Larger seedlings can be also be transplanted, although they require special care. Some trees are more sensitive, others more resistant. Palm trees can be planted when they are three meters high and even higher, without any problem.

The seedlings must be planted in the soil together with the accompanying clod of earth, avoiding bare roots. Cans or plastic bags should be removed when planting. Depending on the size of the seedling and the soil quality, a hole should be dug measuring from 0.50 m x 0.50 m x 0.50 m to 1 m x 1 m x 1 m. It should be prepared beforehand and filled with a mixture of 2/3rds good quality earth and 1/3rd well composted organic manure.

The ideal time for planting, particularly in the summer, is in the early morning or late afternoon, when the sun is not so hot. A wooden or bamboo stake should be used to support and protect the seedling, particularly when planting in public places.

The stake should be driven into the bottom of the hole, leaving about 1.80 m to 2 m above the ground, according to the size of the tree. The seedling is then tied to the stake using a fiber, sisal or rubber strip. Materials which could damage the trunk, like wire, are not recommended.

Irrigation is essential during the first days, and then at longer intervals afterwards. The best way to avoid mistakes is to observe the plant carefully, because each one has different needs, which also change according to the season of the year. The land must be well drained for the water to flow well.

The Reason for Botanical Names

Common names of plants vary according to the country, region and even from one person to another. There are no rules. In contrast, scientific names are controlled by the International Code of Botanical Nomenclature.

The current botanical nomenclature originated from the work of the Swedish botanist Carl von Linné, beginning in the year 1753, when he published *Species plantarum*. Species are classified in Latin, a universal language.

It is not difficult to understand botanical terminology: each species has a name consisting of two words (binomial nomenclature). The first, the genus, is in noun form and always begins with a capital letter. The second, the epithet or specific name, in lowercase letters, is used in adjective form: it describes qualities, origins, facts or persons related in some way to the history of the species. In the case of *Caesalpinia echinata* Lam., for example, *echinata* refers to the prickles on the trunk. The third name is the name of the author who originally described the plant.

Whenever there are more than two names in Latin, the third indicates a specific variety (such as external appearance, color of flower or foliage) of individual plants of the same species. Very often these botanical names are reclassified according to new studies.

Native Trees of Brazil

Anacardium occidentale L.
Family – Anacardiaceae

CASHEW

Characteristics – Mostly medium-sized trees with a full crown. The trunk, which is usually twisted and low, can be tall and almost straight, varying according to the nature of the soil. It can grow up to 20 meters high on fertile soil.

The red or yellow edible fruit are extremely decorative and fragrant and are also an important source of vitamin C.

The cashew nut can be eaten in different ways and has a high commercial value.

Oil is extracted from the hull of the nut for use in making paints, varnishes, paper, insulators and plastics.

The peel of the cashew fruit is astringent. In medicine it is recommended as an antihemorrhagic and for treating ulcers. It is also used as a gargle for mouth sores and throat infections. It contains a reddish substance used for dyeing textiles and for tanning hides.

A resin similar to gum arabic is extracted from the trunk and branches. It provides strong, resistant wood, suitable for civil construction, cabinetmaking, carpentry, for making boxes and farm tool handles. The ashes from its wood contain a high proportion of potassium which can be used as a dentifrice.

Geographical distribution – In fields and on sand dunes all along the northern coast of Brazil.

Anadenanthera peregrina (L) Speg.
Family – Leguminosae

ANGICO

Characteristics – A smooth-trunked tree growing to a height of approximately 15 meters.

It provides hard, heavy wood suitable for civil construction, carpentry, firewood and charcoal. Its bark is used in the tanning industry. It also produces a gum which can be used instead of gum arabic. The Angico is a melliferous tree and produces syrup of excellent quality.

In medicine it is astringent, depurative and hemostatic. It is also used for uterine hemorrhages, for treating leucorrhea and cases of rickets and lack of appetite.

Geographical distribution – From Amazonas and Maranhão down to Paraná and Goiás.

Araucaria angustifolia (Bertol.) O. Kuntze
Family – Araucariaceae

PARANÁ ARAUCARIA or BRAZILIAN PINE

Characteristics – A tall tree whose striking appearance distinguishes it from all other native trees.

Until a few years ago, these pines grew in endless forests which stretched from the Serra de Bocaina as far as the farthest point of Rio Grande do Sul state. For some time this tree was threatened by extinction due, principally, to the exportation of its excellent wood.

Its pine nuts provide large amounts of alcohol and forage. They are also an important source of food for the Indians in the south, who preserve them in different ways.

Geographical distribution – Minas Gerais, São Paulo, Paraná, Santa Catarina and Rio Grande do Sul.

Arecastrum romansoffianum (Mart.)
Family – Palmaceae

QUEEN PALM

Characteristics – This palm sometimes grows to a height of 30 meters.

Its fruit grow in enormous quantities, forming large clusters. As soon as they ripen, they turn golden-yellow, making a beautiful contrast with the crisp, dark green foliage. The pulp of its fruit is succulent, sweet, edible and has an agreeable taste. It is appreciated principally by children and birds and is also used to feed pigs and cattle.

This tree is often used in landscaping schemes. Besides being ornamental, it has the advantage of being easily transplantable, even when the seedlings are quite big.

Geographical distribution – From Ceará down to Rio Grande do Sul, Minas Gerais, Goiás and Mato Grosso.

Barbosa pseudococcus Becc.
Family – Palmaceae

A PALM

Characteristics – A tall palm growing up to 15 meters high. A very decorative tree with a straight, ringed trunk.

Its woody spathe is very heavy. The fruit have a rigid thorn at the end and are not edible. Each nut, however, contains liquid which has a very pleasant taste.

Geographical distribution – Bahia, as far as São Paulo and Minas Gerais.

Basiloxylon brasiliense (Fr.All.) K. Schum.
Family – Sterculiaceae

PAU REI

Characteristics – A majestic, impressive, slender tree. The Pau-rei grows to a height of 20 to 30 meters: its scale is in line with the closeness of large buildings.

Its strong, white wood is much appreciated and was formerly used for making crates for transporting oranges from the northern highlands to England. Besides wood for crates, the pau-rei can provide excellent quality high-yield paper for printing.

The bark is astringent.

It grows preferably in clayey or silico-clayey soils.

Geographical distribution – Northeastern Brazil, principally in Ceará, but also in Espírito Santo and Rio de Janeiro.

Bixa orellana L.
Family – Bixaceae

ANNATTO TREE

Characteristics – A small fast-growing tree.

A very beautiful and decorative tree, not only when it blossoms but also later when it is covered with clusters of colored fruit. The most attractive variety is the Red Annatto, which is widely used for decorating.

Besides its attractive appearance, the Annatto has many uses. The Indians used its dye to paint their ceramics and other vessels for domestic use. They also used it for painting their bodies and, at the same time, to protect their skin against the sun's rays and mosquito bites.

In the Capixaba kitchen no fish casserole is made without Annatto seasoning.

The seeds are used for the most varied industrial purposes, including for making dyes and suntan lotions. Various studies have proved the advantages of using Annatto in milk byproducts, biscuits, soft drinks, wines, meats etc., instead of other dyes of animal origin. Besides adding vitamins A, B and carotene to such products, Annatto is a natural substance which does not have ill effects. Petroleum byproducts used for coloring such foods are considered carcinogenic.

Annatto adapts to any soil and climate, except where there are frosts.

In medicine the seeds are recommended for eye complaints, as antihemorrhagics, aphrodisiacs, laxatives, expectorants and for various heart ailments.

The root contains a substance which aids digestion.

Geographical distribution – In the Amazon region and practically all over Brazil, except in the far southern states.

Bombacopsis glabra (Pasq.)
A. Robyns
Family – Bombacaceae

A SILK-COTTON TREE

Characteristics – A tree with evergreen foliage very often confused with the Pachyra aquatica (another type of silk-cotton tree). The white flowers are sparse and not very abundant, but very striking.

The nuts can be eaten after being cooked, although since they are rich in oil they are somewhat indigestible.

The wood is light and has few uses.

Geographical distribution – Principally in the states of Piauí and Ceará.

Bougainvillea spectabilis Willd.
Hort. tree variety
Family – Nictaginaceae

BOUGAINVILLEA

Characteristics – The only arboreal variety (the others are climbing plants) which has quite straight branches. The trunk can have a diameter of 0.40m to 0.50m.

It is very decorative and colored purple in early spring.

The Bougainvillea genus was dedicated to the French navigator Louis Antoine de Bougainville.

Geographical distribution – The arboreal variety is found in the highlands of Rio de Janeiro state. All the climbing varieties are to be found from Bahia as far as Paraná and Minas Gerais.

Cabralea laevis Dc.
Family – Meliaceae

A CEDAR

Characteristics – A fast-growing tree with large leaves, decorative on account of the shape and arrangement of its leaves.

The wood is used in hydraulic work, in civil construction and ship building and for cabinetmaking. It is also used for wood carving. This tree should not be confused with other types of cedar, particularly the *Cedrela fissilis,* belonging to the mahogany family, much sought after in civil construction because of its color and fragrance.

Geographical distribution – Rio de Janeiro and São Paulo.

Caesalpinia echinata Lamarck
Family – Leguminosae

PRICKLY BRAZILWOOD

Characteristics – This tree may grow 30 meters high, although slowly.

It is a handsome, leafy tree with a thorny trunk and branches, red heart and very hard wood. The golden yellow flowers open from September to December and are sweet smelling.

The Prickly Brazilwood, an almost extinct tree, was one of Brazil's main export products. In a letter dated April 22, 1609, D. Diogo de Menezes wrote to the King of Portugal: "Believe me, Your Majesty, the true mines of Brazil are sugar and the Prickly Brazilwood, which Your Majesty turns to so much account without it costing your farm a single penny."

Its wood is used for civil construction and ship building, as well as in turnery, for making violin bows and other instruments.

An extract is produced from its wood which has medicinal properties as an astringent, tonic and aid for relieving toothache. A good dye for fabrics and fountain pens is also yielded by the Prickly Brazilwood.

Geographical distribution – Along the coast, from Pernambuco down to Rio de Janeiro.

Caesalpinia ferrea Mart.
Family – Leguminosae

IRONWOOD

Characteristics – A tall tree growing up to 30 meters high. The trunk is cylindrical, with smooth bark and large white blotches on a dark background, forming magnificent shapes.

Its yellow blossom appears usually from January to April.

Its fruit has various medicinal properties, principally for treating anemia, hemoptysic ailments, lung diseases and diabetes.

Geographical distribution – A native of the Northeast and Southeast of Brazil, it is found from Ceará down to Rio de Janeiro. It is widely cultivated all over Brazil.

Caesalpinia pelotophoroides Benth
Family – Leguminosae

SIBIPIRUNA

Characteristics – A medium-sized tree.

Often confused with the Prickly Brazilwood, from which it is distinguished by the absence of thorns on the trunk and branches. The Sibipiruna has many branches. When its buds open its foliage is reddish-bronze.

It is highly recommended for planting along streets, avenues, parks and gardens in general, because of its slenderness and abundant yellow blossom which appears from September to November.

Its wood is used for making planks.

Geographical distribution – Widely cultivated in all regions of Brazil. It has been noted growing naturally in forests of Rio de Janeiro, São Paulo and Minas Gerais.

Calycophyllum spruceanum
Hook Fil.
Family – Rubiaceae

MULATTO CALYCOPHYLLUM

Characteristics – A towering, very elegant tree growing up to a height of 16 to 20 meters. Its smooth, glossy trunk only forms a crown very high up. It has magnificent bark the color of old bronze, looking as if it had been varnished. The bark grows again each year. The trunk starts off green and slowly takes on the hue which gave the name to the tree.

Its flowers bloom in the months of May and June.

It produces yellowish-white wood which darkens as it ages, suitable for civil construction, stays, beams and cabinetmaking.

Geographical distribution – A native of the Amazon region.

We suggest a visit to Rio de Janeiro's Botanical Gardens, where there is a magnificent walk bordered by the Mulatto Calcycophyllum.

Cariniana estrellensis Raddi
Family – Lecythidaceae

JEQUITIBÁ

Characteristics – A large tree. In the Center and South of Brazil where it grows up to 40 meters high, it is known as the "giant of the forest".

The trunk is straight and cylindrical and can have a diameter of 2.15 meters. It has no branches until very high up. The foliage is evergreen and the woody fruit hang like thimbles. Its crown is not very wide.

The wood is very hard and very tough to saw.

In medicine its bark is used as an astringent for treating diarrheas and dysenteries. It is also used as a hemostatic. Externally, it is used for gargles, mouth and throat affections and for treating laryngitis and tonsillitis.

Geographical distribution – Rio de Janeiro, Minas Gerais and São Paulo.

Cassia alata L.
Family – Leguminosae

THE RINGWORM SENNA

Characteristics – A small tree or shrub, sometimes growing as high as four meters.

The flowers are large and yellow. The arrangement of the branches and almost black pods remind us of a large candelabra.

It is used in medicine for its diuretic and purgative properties and for fevers. It is also recommended for skin diseases. The root, leaves, bracts and seeds are used.

According to the botanist Pio Corrêa, in certain regions of India, its green seeds are eaten. When roasted they produce an infusion identical to coffee, which is said to strengthen the bowels.

Geographical distribution – From Amazonas to Rio de Janeiro, Minas Gerais, São Paulo, Goiás and Mato Grosso.

Cassia grandis L.
Family – Leguminosae

THE PINK CASSIA

Characteristics – A tall, extremely elegant tree.

At the end of the winter it loses its leaves, although it is in the spring that it attracts attention, when it is covered entirely with lovely, delicate pink flowers, resembling a peach tree.

Geographical distribution – Amazônia and the Pantanal of Mato Grosso.

Cassia macranthera Dc. ex Collad.
Family – Leguminosae

A SENNA

Characteristics – A medium-sized tree. Its rich green, compact foliage contrasts with the yellow of its large, showy flowers which appear from February to April. During this season it combines beautifully with the purple of the glorybushes.

The root is considered an antidote for various poisons, and can also be used to prevent malaria. The leaves are used to make poultices for relieving inflammations.

It is also used for treating rheumatism, asthma, anemias, throat infections and liver ailments. It is a tonic and antiabortive.

Geographical distribution – From Ceará and Rio Grande do Norte down to Santa Catarina and also in Minas Gerais.

Cassia multijuga Rich.
Family – Leguminosae

A SENNA

Characteristics – A medium-sized tree with delicate foliage.

It is notable among secondary association plants because of its exuberant bright yellow blossom which entirely covers its crown, like a lovely bouquet. Its flowers compose and at the same time provide a beautiful contrast with the purple of the *Tibouchinas* (Glorybushes) and the darker yellow of the *Cassia macranthera,* adorning our hillsides all through Lent, or from February to April.

Geographical distribution – São Paulo, Minas Gerais and principally Rio de Janeiro.

Cassia spectabilis excelsea
Schrades
Family – Leguminosae

A SENNA

Characteristics – A medium-sized tree growing approximately seven meters high.

The shape, foliage and particularly the exuberance of its yellow blossom, in September and October, make it a very decorative tree.

Ideal for planting along streets, avenues, in parks and gardens, not only because of its beauty, but also because it adapts easily: it grows as well in the savannah as in regions close to the sea.

Geographical distribution – Savannah and cactus scrub areas of Bahia, Minas Gerais and coastal woodlands, from Ceará down to the south of Brazil.

Cecropia glaziovii Snethlage
Family – Moraceae

SILVERLEAF PUMPWOOD or TRUMPET TREE

Characteristics – A tall slender tree, producing a decorative effect. It can reach a height of 9 to 25 meters, depending on where it grows. Its branches grow at the top, somewhat similar to the Papaya tree.

The trunk is straight and hollow, made up of nodes and internodes and divided into sections. These hollows are always occupied by ants which are housed and fed and, in return, defend their protector against natural predators, that is, any pest or insect which tries to destroy its leaves or flowers. This curious association between the Trumpet tree and the ant is known as "mutualism".

The enormous leaves are sometimes a meter long and have a handsome undulating shape. Their silvery hue is always eye-catching, making a lovely contrast with the many shades of green of our forests. They are so abrasive that they are used as sandpaper to polish wood.

The Trumpet tree flowers and bears fruit in the summer. The seeds are much relished by birds and particularly by the sloth, which feeds on its flower clusters and has even loaned its name to the tree, sometimes called the "sloth tree".

The bark provides strong fibers. The trunk is used for making boxes and matches and for building rafts and floats. The Carib Indians used the hollow trunk to make musical instruments. In the interior of Minas Gerais young trunks used to be made into roof rainwater gutters.

The shoots are the parts used medicinally as an astringent for treating the respiratory system, hemorrhages, asthma, bronchitis and coughs, as well as dysentery and diarrhea. It increases the energy of the heart muscle, causing greater diuresis.

Geographical distribution – Practically all over Brazil, in the more humid, hot regions.

Ceiba petandra (L.) Gaert
Family – Bombacaceae

KAPOK CEIBA

Characteristics – Considered one of the largest trees of the tropical forest, it can grow more than 50 meters high and live over a hundred years. It grows very quickly in hot climates. Its tubular roots, or prop roots, besides their ornamental effect, are true niches which the Indians turn into homes. They consider it the "mother of trees".

The floss of the Kapok Ceiba, which is of excellent quality, has many uses and is preferred for making lifejackets and lifebelts, due to its remarkable flexibility. It is also a valuable heat insulator and withstands 30 to 35 times its weight in water.

This tree's flowers have a slightly cheesy smell and its wood can be used for boat building and cellulose for making paper. The seeds are oilbearing and edible.

In folklore remedies, the bark and sap are used for treating conjunctivitis, while in medicine they are used for treating skin injuries.

Geographical distribution – Amazônia.

Chorisia crispiflora H.B.K.
Family – Bombacaceae

THE FLOSS-SILK TREE

Characteristics – A colorful tree reaching a height of 16 to 20 meters. The trunk is slightly potbellied and covered with short, fat thorns.

It sheds all its leaves in March and April, when it is covered with large pink flowers giving it a magnificent appearance. It is also very beautiful in early spring, when its bolls open to release the abundant, pale floss-silk.

Besides its floss, which is used for making pillows, quilting, etc., its white, very light and porous wood can be used forbuilding canoes and paper making.

Geographical distribution – Rio de Janeiro, São Paulo and Paraná, down to Santa Catarina.

Citharexylum myrianthum Cham.
Family – Verbenaceae

A FIDDLEWOOD

Characteristics – A fast-growing medium-sized tree. As soon as it sheds its leaves, white sweet-smelling flowers appear. After it has blossomed, it is covered with clusters of small, shiny red fruit which are very decorative and attractive to birds.

The wood is very resistant to damp and suitable for carpentry, box making, pulp for making paper and musical instruments such as guitars, hence one of the names by which it is known: Fiddlewood.

Geographical distribution – From Amazonas to São Paulo.

Clitoria fairchildiana Howard
Family – Leguminosae

SOMBREIRO

Characteristics – A tree of average height.

Its broad, leafy crown is excellent for giving shade, hence its common name. It sheds all its leaves in July and August, although they sprout again in early spring.

It grows very fast. Its yellowish-white wood is not very strong. The purplish flowers are abundant and open out into fairly long clusters between January and March.

Besides parks and woods, it is recommended mainly for planting along roadsides and for shading pastures.

Geographical distribution – Amazon region.

Couroupita guianensis Aublet
Family – Lecythidaceae

THE GUIANA CANNONBALL TREE

Characteristics – A tall, majestic tree sometimes growing as much as 20 meters high.

A bizarre tree, it is made singularly beautiful by its countless inflorescences growing from the trunk. It flowers for a long period, usually from October to March. At this time it resembles a large column decorated and surrounded by large, strikingly lovely red flowers. Their strong fragrance can be noted from quite a distance.

The large globular fruit are appreciated by monkeys.

The wood is soft and has little value, although it can be used for making boxes.

It is suitable for parks, squares and gardens, although not for street planting, parking zones and pedestrian areas, because its fruit are very heavy and can cause accidents when they fall from high up.

Geographical distribution – Amazon region.

Crescentia cujete L.
Family – Bignoneaceae

COMMON CALABASHTREE

Characteristics – A stocky, leafy tree with a twisted trunk.

It is very graceful and decorative when it curves the weight of its large and abundant fruit. A raw black substance is extracted from the calabash, its fruit, which is used for dyeing fabrics. The calabash can also be used in the home, apart from being decorative.

The seeds are much appreciated by the Mexicans and can be eaten boiled or roasted.

The wood is good for cabinetmaking, carpentry and making instrument handles, although it has the drawback of rotting very fast when in contact with damp.

The pulp of the green fruit of the calabash has several medicinal properties: mixed with honey and sugar it is an expectorant; it is also a purgative and reduces fever. The ripe pulp is used in poultices as an emollient.

Geographical distribution – From Amazônia to Rio de Janeiro and Goiás.

Erythrina crista-galli L.
Family – Leguminosae

COCKSPUR CORALBEAN

Characteristics – A small tree growing from seven to ten meters high.

Its trunk is generally twisted and covered with a thick layer of corky bark, which is why it is also known as the cork tree. These attributes make it an excellent host for orchids which find its trunk a suitable place to fasten themselves.

The Cockspur Coralbean prefers rather wet soil, although it adapts well to other types. Its crown is widely spread out and its branches when new are full of thorns. The flowers are scarlet.

Geographical distribution – Rio de Janeiro, Minas Gerais, São Paulo, Paraná, Santa Catarina and Rio Grande do Sul.

Erythrina falcata Benth.
Family – Leguminosae

A CORALBEAN

Characteristics – This tree grows up to 20 meters high.

Almost all the Erythrinas flower in the winter months. The *falcata* is the first to shed its leaves, covering itself with clusters of brick-red flowers.

In certain regions, soon after it blossoms, *Maracanãs* (small macaws) appear, birds rather similar to Maritaca parrots, also green but larger. The *Maracanãs* live in small flocks, while the *Maritacas* are always seen in large groups.

This tree tolerates drier regions well. Its wood can be used for making ceilings, laths or matches.

Geographical distribution – Minas Gerais, São Paulo, Rio de Janeiro, Paraná, Santa Catarina and Rio Grande do Sul.

Erythrina fluminensis Barneby & Krukoff
Family – Leguminosae

A CORALBEAN (FLUMINENSIS)

Characteristics – A robust, medium-sized tree. A hybrid much appreciated for landscape gardening. It is covered with salmon-colored flowers in the winter months.

Geographical distribution – Discovered in the Botanical Nursery of the National Museum in Rio de Janeiro and cultivated in Flamengo Park.

Erythrina fusca Loureiro
Family – Leguminosae

A CORALBEAN (FUSCA)

Characteristics – A large tree growing up to 20 meters high.

Its fast growth makes it an ideal tree for parks. It adapts to very varying regions: it grows as well in river lowlands as on dry prairies.

Its salmon pink flowers bloom in the winter.

The branches are brittle. Its light wood is used for making rafts. It is also much used for shading coffee and cocoa plantations.

It is used in medicine for its antirheumatic properties.

Geographical distribution – Amazônia.

Erythrina speciosa Andrews
Family – Leguminosae

A CORALBEAN

Characteristics – A small very thorny tree with many branches.

In the winter it sheds all its leaves, displaying its bare branches shaped like candelabra. In July and September the tips of its branches are full of clusters of bright red flowers.

It attracts hummingbirds and is an excellent host for all kinds of orchids.

It adapts well to any climate and is very resistant to prolonged droughts, although it can grow in damp soil.

Geographical distribution – Minas Gerais, Rio de Janeiro, São Paulo, Paraná, Bahia, Espírito Santo and Santa Catarina, principally in coastal woodlands.

Gallesia gorazema Mok.
Family – Phytolaccaceae

GUARAREMA

Characteristics – A large tree growing up to 38 meters high. The twisted and usually hollow trunk can measure up to 3 meters in diameter.

It provides offwhite, light, soft and easily worked lumber. Similar to the other parts of this tree, when it is freshly sawed it exhales a strong smell resembling garlic. For this reason, in the countryside, they say that snakes do not go near places where there is a Guararema. Its presence is a sign of fertile soil.

In folk medicine it is used as an antispasmodic and for relieving nervous coughs. The wood, ground up and mixed with the leaves and flowers is used in baths for rheumatism.

Geographical distribution – Rio de Janeiro, São Paulo and Minas Gerais.

Gustavia augusta L.
Family – Lecythidaceae

JENIPARANÁ

Characteristics – A large tree growing up to 13 meters high.

It is excellent for providing shade. The pink flowers are sweet-smelling and very attractive.

It produces hard, resistant wood recommended for civil construction and cabinetmaking.

Its bitter root has therapeutic properties for liver ailments, including jaundice. The leaves are decongestant when used in poultices.

Geographical distribution – It is found in the wetlands of Amazônia, but adapts well in other regions.

Holocalyx glaziowii Taub.
Family – Leguminosae

ALECRIM DE CAMPINAS

Characteristics – A medium-sized tree with fine, smooth bark.

Its delicate perennial foliage is a shiny dark green. Its white flowers grow in small clusters, although they are not particularly decorative.

The wood is very resistant and suitable for cabinetmaking, turnery, and for making the shafts and spokes of wheels. The Xavante Indians use it for making clubs and the tips of arrows.

This tree is widely used for planting alongside roads throughout São Paulo state and particularly in Campinas, from where its best-known name comes. It tolerates pruning and withstands frosts.

The bark and leaves have therapeutic properties, including for reducing fevers.

Geographical distribution – From São Paulo to Rio Grande do Sul.

Hymenea courbaril L.
Family – Leguminosae

THE COURBARIL TREE

Characteristics – A large tree which can grow more than 20 meters high. It grows fast and it has deep taproots. A lofty and impressive tree, with a straight, smooth, pinkish-brown trunk.

The bark exudes a resinous balsam commercially known as copal, used for making varnishes and lacquers.

The fruit are thick pods which drop off whole, containing three to five seeds enclosed inside a mealy, sweet and edible pulp.

The foliage is evergreen and forms a rounded crown, giving medium shade. It has cream-colored flowers. These features make it recommendable for parks, squares and gardens.

The wood is hard, attractive and appreciated for cabinetmaking.

The pulp of the fruit, trunk, branches and bark have various therapeutic properties as an expectorant, astringent, chest decongestant, arterial sedative, digestive stimulant, tonic and balsamic. It is also recommended for lung ailments and laryngitis.

Geographical distribution – From Amazônia to São Paulo, also Minas Gerais and Mato Grosso.

Inga Sp.
Family – Leguminosae

INGÁ

Characteristics – Amongst trees and shrubs, more then 200 species belonging to this family are known and have been classified. The wood of most of them is white or reddish.

Generally, *Ingazeiros* grow along the banks of rivers and reservoirs. The seeds falling from the trees are appreciated by fish, although they can also be eaten by man.

It is a melliferous tree. In the Amazon region the Indians celebrate the appearance of the ripe fruit by making a beverage of them similar to wine.

The Ingá is recommended for protecting land against erosion and also for shading coffee plantations.

It is a very ornamental tree.

Geographical distribution – Principally in the Amazon region and from Ceará to Rio Grande do Sul, including in Minas Gerais.

Jacaranda mimosaefolia D. Don
Family – Bignoniaceae

THE SHARPLEAF JACARANDA

Characteristics – In spite of its name, it is not related to trees of the Leguminosae family which produce hardwood. Only the native origin of the common name is shared by the two species: *iá-acarantá*, which means wood which has a hard pith.

A medium-sized tree, it grows up to 8 meters high. It is extremely decorative. The delicate foliage, highlighted by the exquisite violet-blue blossom, attracts countless hummingbirds. The sweet-smelling flowers open from September onward.

Recommended for planting along streets and in parks, squares and gardens.

As a medicinal plant it is indicated for painful joints and rheumatism. It is considered an antisyphilitic, depurative, sudorific and diuretic. The leaves are used for medicinal purposes. A powder made from them is used externally for healing wounds.

Geographical distribution – Mainly in Central Brazil.

Lafoensia glyptocarpa Koehne
Family – Lythraceae

MIRINDIBA

Characteristics – A large, elegant tree with glossy foliage.

Between July and August it is covered with white flowers studded with countless red buttons.

Suitable for live hedges, planting along streets, in parks, squares and for reforestation.

Its hard wood is used in civil construction, carpentry and cabinet-making. It is also used for producing firewood.

Geographical distribution – Rio de Janeiro, Espírito Santo and Minas Gerais.

Lecythis pisonis Camb.
Family – Lecythidaceae

SAPUCAIA-NUT TREE

Characteristics – A large, fast-growing tree. It grows 20 to 30 meters high and its trunk can have a diameter of 1 meter.

The crown is broad. During a short period it sheds all its leaves, followed by extraordinarily beautiful pink foliage which is its greatest attraction. Almost at the same time white flowers open up, although they are not very eye-catching.

The fruit of the *Sapucaia* — shaped like a gourd bottle — are much appreciated by monkeys which like their seeds. Usually they manage to grasp the seeds inside the woody capsule, but when they try to remove their hands they get stuck or hurt themselves. This is the origin of the proverb "the wise monkey doesn't put his hand in a gourd".

Its wood is heavy and hard, strong and very durable. It is used in civil construction and for boat building, railroad sleepers, struts, beams, etc.

The bark is used medicinally for relieving coughs.

Geographical distribution – From Ceará to Rio de Janeiro, particularly in the south of Bahia and north of Espírito Santo states.

Licania tomentosa Benth.
Family – Chrysobalanaceae

OITI

Characteristics – A leafy tree growing up to 15 meters high.

The trunk is straight and covered by dark brown bark.

The crown is thick and compact and consists of evergreen foliage. These two features and the fact that it stands up well to adverse urban conditions, make it a tree recommended for planting along streets.

The flowers are not remarkable, but the leaves when new form an all-white crown giving it a very attractive appearance.

The fruit are edible and the nuts are very rich in oil.

Its wood is suitable for civil construction and hydraulic uses.

Geographical distribution – From Piauí to Bahia.

Lophantera lactescens Ducke
Family – Malpighiaceae

LANTERNEIRA

Characteristics – A slender, small to medium-sized tree with a straight trunk. It grows quickly.

The leaves are large and decorative. From August to September its yellow flowers form long clusters 0.20m to 0.30m long, giving it an extremely beautiful and decorative appearance.

Highly recommended for gardens, parks and avenues.

Although a native of Amazônia, it adapts well to other climates, except very cold regions.

Geographical distribution – Amazônia.

Mimosa Schonburghii Benth.
Family – Leguminosae

A MIMOSA

Characteristics – A small to medium-sized tree growing 8 to 10 meters high.

The first blossom appears early on, when it is only 2.5 meters high. The white flowers entirely cover the crown, giving it a very decorative appearance.

Geographical distribution – Amazonas and Rio de Janeiro state.

Myrciaria cauliflora Berg.
Family – Myrtaceae

JABOTICABA

Characteristics – A medium-sized slow-growing tree with a full crown.

The round, shiny, dark purple fruit are scattered all over its trunks and branches, making it extremely beautiful. Due to their delicious flavor, they are much sought by man and birds. They can be made into wine, preserves and liqueurs, among other products.

Due to its strength, the wood is used for internal construction, the framework of beams and struts.

Jaboticaba bark is astringent and is used for treating asthma, dysentery, erysipelas and inflammations of the tonsils. It is also used for making an extract for coloring wines, vinegars, etc.

Geographical distribution – Native all over Brazil, although it is more common in Minas Gerais, Espírito Santo, Rio de Janeiro, São Paulo and Paraná states.

Pachyra aquatica Aublet
Family – Bombacaceae

A SILK-COTTON TREE

Characteristics – Pachira is a native name from the Guianas and means preference for humid places.

The leaves of this silk-cotton tree are emollient and edible. The seeds, which are much liked by monkeys, can be eaten boiled or roasted, and are excellent for fattening pigs to whose meat they impart an agreeable flavor. Formerly, these seeds were wrongly known as "Coast of Spain nuts".

Although it prefers damp places, it adapts well to street-planting. Its sweet-smelling flowers are like large white sponges with many red stamens.

Geographical distribution – Amazônia. It is often used for planting along streets in Rio de Janeiro, where it is known as the *Munguba*.

Peltophorum dubium Taub.
Family – Leguminosae

A PELTOPHORUM

Characteristics – A tall tree with a broad crown.

The light and delicate foliage and yellow blossom, appearing from October to February, makes it a very ornamental tree.

Its wood marked with light or dark stripes or blotches, is suitable for civil construction, cabinetmaking and for making dyes.

The bark contains tannin and is used for tanning hides.

The roots, fruit and leaves are used in folk medicine.

Geographical distribution – Bahia, Minas Gerais, Rio de Janeiro, São Paulo and Mato Grosso.

Piptadenia gonoacantha Mart.
Family – Leguminosae

JACARÉ

Characteristics – A fast-growing tree. The bark and branches have ridges like those on an alligator's (jacaré) back, which is the reason for its common name.

The wood has a soft, white pith and is suitable for firewood and charcoal, burning even when it is green. It can also be used for indoor construction and struts.

This tree can be used for shading coffee plantations and its flowers are valuable as a source of honey.

It is easy to grow, although its seeds should be sown immediately after harvesting, since their germination period is short.

Geographical distribution – From the north to the south of Brazil.

Pterocarpus rohrii Mich.
Family—Leguminosae

A PADAUK TREE

Characteristics—This tree grows from 15 to 20 meters high.

The fragrant flowers are yellow. One of the petals has a violet streak on it.

Its wood is white and soft and makes good pulp for paper making.

Its cellulose yield is excellent.

Geographical distribution—From Amazonas and Pará, all along the coast as far down as Santa Catarina, including Minas Gerais.

Pithecellobium glaziovii
Bentham
Family—Leguminosae

AN APE'S-EARRING TREE

Characteristics—An imposing tree.

It blossoms from September to December, producing small white flowers exhaling a very pleasing scent, which are much sought by bees.

The fruit resembles a spirally rolled pod and is red inside.

A very decorative tree recommended for squares and parks, due to its broad crown.

Geographical distribution—Forests of Rio de Janeiro.

Schinus terebinthifolius Raddi
Family—Anacardiaceae

BRAZIL PEPPERTREE

Characteristics—A medium-sized, extremely sturdy tree.

Its many flowers are white but not eye-catching, producing a profusion of small red fruit which are much appreciated by birds and make it very attractive.

The branches of this tree when new are used for whitening and cleaning teeth. The bark is used for tanning hides and strengthening fishing nets.

Its wood is dark and soft, although heavy and very strong. It can be used for making struts, fence posts, firewood and charcoal.

It is recommended medicinally as an astringent. Its bark is considered depurative and is used for bringing down fevers, for respiratory expectoration and for treating diarrhea. It is also recommended as a balsam for diseases of the urinary system and for lymphatic tumours. Besides the bark, the leaves and fruit are also used.

When it was taken to Europe it adapted to the Côte D'Azur, where it may be found in parks and along avenues.

Geographical distribution—Bahia and Minas Gerais, down as far as Rio Grande do Sul.

Schizolobium parahyba (Vell.) Blake
Family—Leguminosae

A SENNA

Characteristics—An elegant, majestic tree, growing up to 30 meters high. The smooth trunk often has a diameter of more than a meter.

It grows incredibly fast and is one of the tallest trees of the Serra do Mar mountain range. The immense leaves are nonetheless delicate in appearance, resembling the leaves of the dicksonia fern.

It sheds all its leaves in the winter months. Its yellow flowers open from September to November in the form of bouquets.

Folklore would have it that its seeds ward off the evil eye. They can also be used for gambling chips; hence its local name of "ficheira" in some places. The bark is used by tanneries.

Its light, white wood is suitable for making boats, ceiling boards, clogs, matches and principally for the paper industry.

It is also suitable for parks and wide open spaces. It is not recommended for planting near constructions, because its branches break easily in strong winds.

The seeds and bark have astringent, stimulant, stomachic, tonic and aromatic properties and are also used for relieving toothache.

Geographical distribution—Coastal woodlands, from Bahia down to Paraná.

Stifftia chrysantha Mikan
Family—Compositae

THE BRAZIL STIFFTIA

Characteristics—A small, very decorative tree. Its attractive yellow flowers last a long time and are widely used by florists for flower arrangements. It is also recommended for gardens and parks.

Besides this species, there is another rarer type which has red flowers and whose first seeds were collected on the Serra do Mar mountain range. It was named *Stifftia graziella*, by Prof. Hermógenes Freitas Leitão, as a tribute to botanist Graziella Barroso.

There is also another species called *Stifftia parviflora* Dou, producing white flowers, found in Pinhal municipality, State of São Paulo.

Geographical distribution—From Bahia to São Paulo, principally in the woodlands of Rio de Janeiro state.

Tabebuia chrysotricha Mart.
Family—Bignoniaceae

THE YELLOW TECOMA

Characteristics—The scientific name of this tree, *Tabebuia*, comes from the Tupi-Guarani language and means "that which floats".

A medium-sized tree with a straight trunk and dusky bark.

It sheds all its leaves in the winter offering us a magnificent sight of spectacular golden-yellow blossom.

Its wood is very hard. The inner bark has therapeutic properties as an astringent. It is used for treating throat ailments and stomatitis.

Geographical Distribution—Rio de Janeiro, São Paulo and Minas Gerais.

Tabebuia heptaphylla
(Vell.) Mart.
Family – Bignoniaceae

THE PURPLE TECOMA

Characteristics – A tree growing from 10 to 20 meters high, with a round trunk and grayish bark.

It is the first of the Tecomas to flower. Its purplish-red blossoms open as soon as it sheds its leaves from June to August, giving it a very ornamental appearance. Its full glory lasts an average of 15 days. As soon as the flowers fall, the leaves sprout.

The bark on the trunk is recommended for preventing the growth of malignant tumours.

Geographical distribution – Center and South of Brazil.

Tabebuia odontodiscus
Bur. Schum.
Family – Bignoniaceae

THE WHITE TECOMA

Characteristics – A medium-sized tree which prefers tropical climates. It does not flower in regions where there are low temperatures in the winter.

Its blossom from September to October is really spectacular. The only drawback is the short time it lasts. It is still worth it as a simply dazzling sight.

Geographical distribution – States of Rio de Janeiro and Minas Gerais.

Tibouchina granulosa Cogan.
Family – Melastomataceae

THE PURPLE GLORYBUSH

Characteristics – A small tree growing 5 to 12 meters high. It grows fast and has evergreen foliage.

The very decorative purple blossom is always abundant and lasts from February to April. Its purple flowers provide a lovely contrast with the yellow bouquets of the *Cassias macrantha* and *multijuga*, which bloom at the same time.

Geographical distribution – Rio de Janeiro, São Paulo, Minas Gerais, Bahia and Pará.

Tibouchina stenocarpa Cogan.
Family – Melastomataceae

THE PINK GLORYBUSH

Characteristics – A small tree with evergreen foliage.

From February to April it turns into a true bouquet of very beautiful pink flowers.

It is the rarest of the glorybushes. Its seeds very seldom produce the same variety and it should therefore be propagated by using slightly woody stakes or layering.

Geographical distribution – Rio de Janeiro, São Paulo and Minas Gerais.

Tibouchina mutabillis Cogan.
Family – Melastomataceae

YESTERDAY-TODAY-AND-TOMORROW

Characteristics – A shrub with evergreen foliage.

The flowers bloom white and gradually turn from lilac to deep purple. Due to this variation it is somewhat similar to the Manaca Raintree (*Brunfelsia*), which adorns gardens with its lovely, sweet-smelling flowers.

Geographical distribution – Rio de Janeiro and São Paulo. It is very often found in the Santos highlands and along the highway between São Paulo and Paraná.

Triplaris surinamensis Cham.
Family – Polygonaceae

AN ANT TREE

Characteristics – A tall, slender tree with a straight trunk, reaching a height of 10 to 15 meters.

Its crown is small and conical, in a pyramid shape. The leaves are large. When it blossoms for many days between April and June, it becomes very decorative.

It shelters ant colonies inside its boughs. They communicate through true "corridors", but do not harm the tree, cohabiting with it in a state of symbiosis.

The wood is suitable for civil construction, cabinetmaking and carpentry.

Geographical distribution - Amazônia.

Acknowledgements

My thanks:

To José Carlos, my husband and dedicated companion while searching for and gathering new plants. During our trips, he documented with his camera the fine species we found. Some of those photos have been included in this book.

To Professor Luiz Emygdio de Mello Filho, to whom I owe so much and with whom I learned most, on countless occasions resorting to this true walking botanical encyclopedia. Always promptly answering all my questions, his help was of priceless value for the botanical classification.

To Graziella Barroso and her team, who helped me tremendously, identifying my "discoveries".

To Liana Fortes, who supervised the text, whom I often consulted about including passages or modifying others, so as to add to the information.

To Marly and Haroldo Lima, who guided my efforts in the botanical classification of the leguminous plants.

To my daughter Lídia, for her infinite patience in typing my "scribbling".

To Paula Laclette, museum specialist, who so kindly loaned me photos from her collection.

To Carmen Moretzsohn Rocha, Librarian of the Brazilian Nature Conservation Foundation, for the attention she gave me in helping with my bibliographical references.

To my publishers, Geraldo Jordão Pereira and his son Marcos, who made this book possible, my dream of many years.

To my friends for all their encouragement, and to all those who helped me in some way or another and whom I would not like to forget. To all those who sowed, planted and harvested with me, specially Moisés Neto da Silva and Sebastião Otero, who are with us until today.

I would also like to pay a special tribute to a great human being who is no longer among us: Dimitri Lambru, with whom I first planned this book.

Cecilia Beatriz da Veiga Soares

Index of Native Trees

Bibliography

BLOSSFELD, Harry. Jardinagem. – São Paulo: Edições Melhoramentos, 1965 - 418 p.

CORREA, Pio M. – Dicionário de plantas úteis do Brasil - Rio de Janeiro: Ministry of Agriculture/ IBDF. 1975. 6 Vol.

COIMBRA, Raul. – Notas de Fitoterapia - 2nd ed. Rio de Janeiro: Editora Silva Araújo, 1958. 427 p.

RIZZINI, Carlos T. & MOHRS, Walter B. – Botânica Econômica Brasileira, - São Paulo: São Paulo University Press, 1976. 207 ill. p.

WETSTEIN, Richard R.V. – Plantas do Brasil: Aspectos da vegetação do sul do Brasil: Translated by Bertha Lange do Morretes: São Paulo: Ed. Edgard Blücher. São Paulo University Press, 1970, 126 ill. p. Color.

SOUZA, Hermes Moreira de. – [Miscellaneous articles] O Estado de São Paulo, Agricultural Supplement, São Paulo (1334, 1335, 1375, 1396 and 1420): 1981 to 1982.

ÁLBUM Florístico. – Rio de Janeiro: Ministry of Agriculture, Forestry Service, 1945. ill.

FLORESTA da Tijuca/ATALA Fuad et. al. Rio de Janeiro: Nature Conservation Center, 1966, 152 p.

ARBORETO Carioca. – Rio de Janeiro: Nature Conservation Center, 1963/ 1969. 4 Vol. ill.

REFLORESTAMENTO Heterogêneo com essências indígenas – Boletim técnico, Forestry Institute, São Paulo (24), March, 1977.

MELLO FILHO, Luiz Emygdio. Introdução no estudo do gênero, Ficus (Moraceae) na Guanabara e Arredores – Rio de Janeiro, 1963 - Thesis for competitive examination of University Professor (Botany) of Guanabara State University.

INVENTÁRIO de Plantas Medicinais do Estado da Bahia/Coordinated by SILVA, Maria Conceição da. Under-Secretary of Science & Technology. – Salvador: 1979. 1206 p.

FORESTRY SERVICE ARCHIVES. – Vol. 6 - Rio de Janeiro - Ministry of Agriculture, 1952 - 200 p.

O KORUPIRA. FBCN Bulletin, 5: 32-53, 1970.